Name _____

ACTIVITIES FOR SOLAR SYSTEM

Hint: To help you remember the planets in their correct order from the sun, memorize this sentence.

My	**V**ery	**E**ducated	**M**other	**J**ust	**S**erved	**U**s	**N**achos
e	e	a	a	u	a	r	e
r	n	r	r	p	t	a	p
c	u	t	s	i	u	n	t
u	s	h		t	r	u	u
r				e	n	s	n
y				r			e

Write the answers on the blanks.

1. The Sun is the center of what system? _____
2. What are comets? _____
3. Explain the main difference between a planet and a star. _____
_____
_____

4. Name the largest member of our Solar System. _____
5. Does the Sun travel around the planets or do the planets travel around the Sun?
_____

6. Our Solar System is in what Galaxy? _____
7. How does our Solar System compare in size to this Galaxy? _____
_____

8. Fill in the names of the planets in their correct order in space.

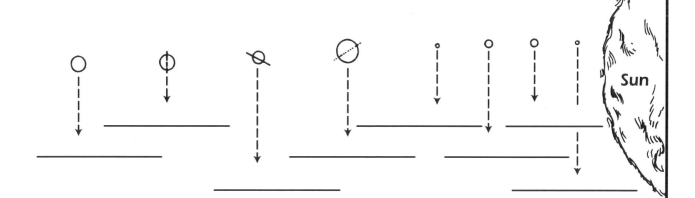

MP3409 Solar System

3

Copyright © Milliken Publishing Co. All rights reserved.

Orbits

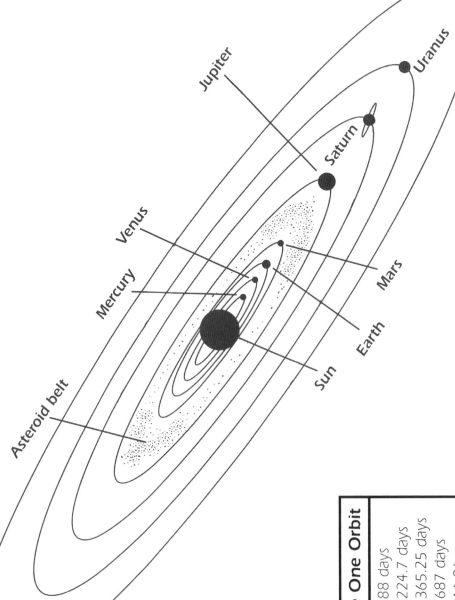

Time to Complete One Orbit	
Mercury	88 days
Venus	224.7 days
Earth	365.25 days
Mars	687 days
Jupiter	11.86 years
Saturn	29.5 years
Uranus	84 Years
Neptune	164.75 years

Name _____ Date _____

SOLAR SYSTEM

OUR SOLAR SYSTEM

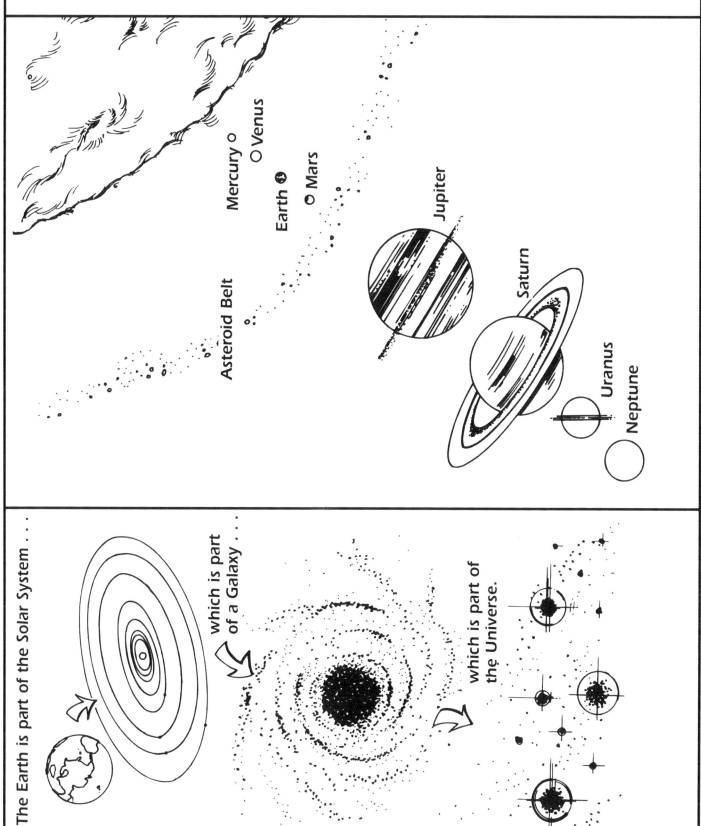

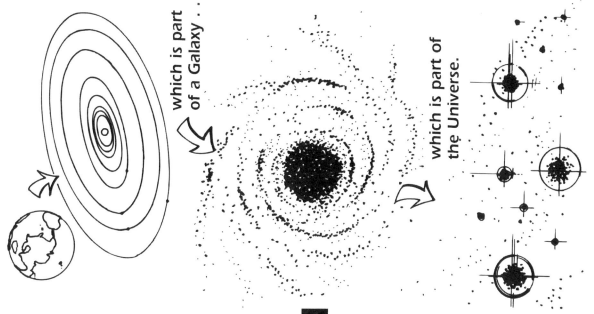

MP3409 Solar System

Our Solar System

The Solar System is made up of the Sun and its family of eight planets. Pluto, formerly the ninth planet, was reclassified as a dwarf planet in August of 2006 when similar-sized objects were discovered. Scientists decided to finally define "planet" when they found Eris to be larger than Pluto. Orbiting between the planets are chunks of rock and metal called **asteroids**, as well as loose collections of rock and frozen gas called **comets**.

Planets and stars are different from one another. A planet is a body that does not give off light of its own. It reflects the light of the Sun. Stars produce and give off their own heat and light. Our Sun is a star.

Our Sun is the center of our Solar System and the planets revolve around it. It is also the largest member. The word solar means "of the Sun."

Wherever the Sun goes in space, the rest of the family or system goes too.

The Solar System is part of an even larger family in space. The Sun is only one of a huge group of stars called the Milky Way Galaxy. Compared to the Milky Way Galaxy, our Solar System is an extremely tiny speck. There are billions of stars in this Galaxy and our Sun is only one.

Our study does not end there. If you look through a powerful telescope, you would see millions and millions of other galaxies, each containing billions of stars of their own. Our Solar System, the Milky Way, and all the other galaxies make up the Universe. Everything in space is part of the Universe.

Solar System

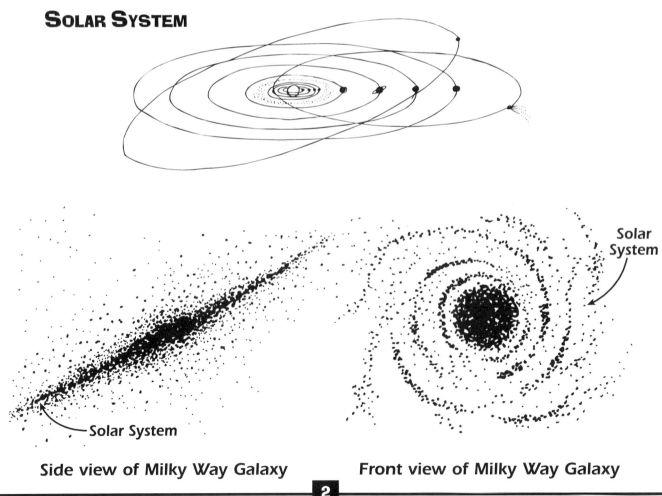

Side view of Milky Way Galaxy

Front view of Milky Way Galaxy

Name _____   Date _____

SOLAR SYSTEM

ORBITS

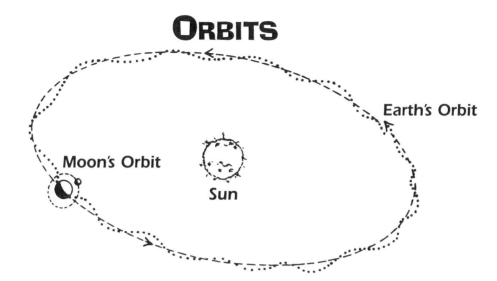

The word **orbit** describes the curved path that a body follows when revolving around another body. The moon travels around the Earth in an orbit. The planets travel around the Sun in orbits. You can also use the word orbit as a verb. The moon orbits the Earth. The planets orbit the Sun. The path of their orbit is an ellipse. An ellipse may be nearly a circle or it may be very elongated.

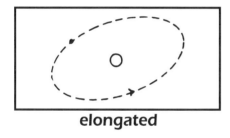
elongated

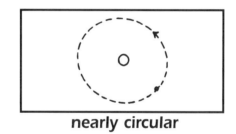
nearly circular

DRAWING ELLIPSES

MATERIALS

two thumbtacks   24" string   unlined paper
pencil   8½ x 11" corrugated cardboard

ACTIVITY

1. Place a sheet of paper on the cardboard. Push two thumbtacks into the middle of the paper about two to three inches apart. Make a small loop out of eight to ten inches of string and put it around the tacks. Pull the loop of string tight with the pencil. Draw an ellipse as you move the pencil around.

2. Vary the position of the tacks and the length of the string to make different shaped ellipses. All planetary orbits are ellipses (very slight) and not circular as most people think.

EXTRA CREDIT

Write a report telling who Johannes Kepler was and why he is important in our study of the Solar System.

Name _____   Date _____

SOLAR SYSTEM

SOLAR SYSTEM AND ORBIT REVIEW

Write **T** for true or **F** for false on the blanks before each sentence. If the sentence is false, rewrite it to make it true.

1. _____ A planet is a body that gives off light of its own. _____
   _____

2. _____ The Sun is the largest member of our Solar System. _____
   _____

3. _____ The Universe is made up of only our Solar System. _____
   _____

4. _____ The Earth orbits the moon. _____
   _____

5. _____ Neptune has the fastest orbit around the Sun. _____
   _____

6. _____ The planets orbit the Sun. _____
   _____

7. _____ The Earth takes one year or 365.25 days to orbit the moon. _____
   _____

8. _____ The Sun is the center of our Solar System. _____
   _____

9. _____ There are eight planets in our Solar System. _____
   _____

10. _____ The correct order of the planets from the Sun is: Mercury, Venus, Earth, Mars, Jupiter, Uranus, Neptune, Saturn. _____
    _____

11. _____ A star gives off heat and light of its own.
    _____
    _____
    _____

Name _____   Date _____

SOLAR SYSTEM

THE SUN

The Sun is a star made up of hot gases that explode with energy similar to that of a continuously exploding nuclear bomb. It is the center of our Solar System. It provides us with heat and light. The Sun has been spinning on its axis and exploding for about 5 billion years.

The Sun is an average-size star, but seems larger because it is the star nearest to us—only 93,000,000 miles (150,000,000 km) away. This is a very, very long way, but the other stars are even farther out in space. Light from the Sun takes about eight minutes to reach us, so actually we see the Sun as it was eight minutes ago!

The **corona** is the outer part of the Sun's atmosphere. The **chromosphere** is made of very, very hot gases which shoot up into the corona at high speeds. Heat is sent to the surface of the Sun through the **middle** and **outer** layers from the **core**. The temperature of the core is approximately 57,000,000°F (31,350,000°C).

inside the Sun

Label the diagram of the sun.

A. _____
B. _____
C. _____
D. _____
E. _____

The Sun is much larger than the Earth. The diameter (distance across) of the Sun is 109 times that of the Earth. For comparison, you could fit about one million Earths inside the Sun!

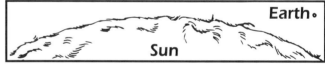

Sun's size compared to the Earth

MAKE A SOLAR COLLECTOR

MATERIALS
2 cans the same size   black paper   thermometer

ACTIVITY
1. Cover one can with black paper. Fill both cans with water.
2. Set both cans outside in the sun on a warm sunny day.
3. Record the water temperature at the beginning of the experiment and again after 10, 20, and 30 minutes.
4. Which can collected the most solar energy?

MP3409 Solar System   Copyright © Milliken Publishing Co. All rights reserved.

SOLAR SYSTEM

The Sun

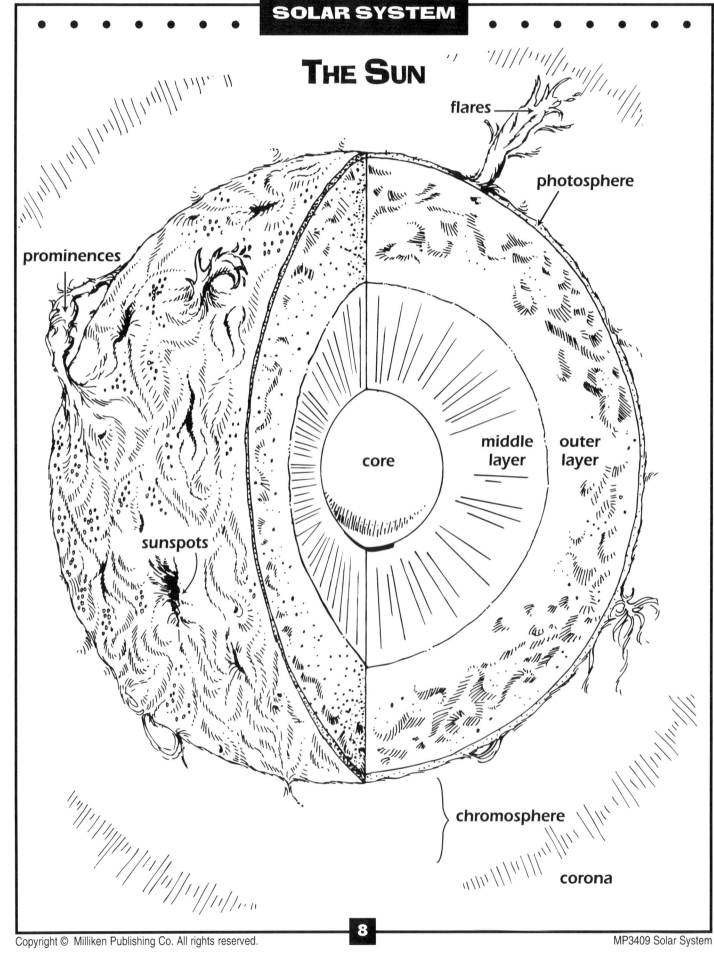

SOLAR SYSTEM

HOW THE SUN PRODUCES ENERGY

Define these words. Use a dictionary if you need help.

1. energy _____
2. atom _____
3. hydrogen _____
4. helium _____

Deep within the core, there are atoms of hydrogen that are under pressure and are turning into atoms of helium. As they do this, energy is given off and sent out into space as heat energy and light energy. Because this energy is being produced deep within the Sun, it takes millions of years for it to reach the surface and escape into space.

Less than one percent of the Sun's energy reaches Earth. Without this energy, there would be no life on our planet.

On the surface of the Sun, there are dark spots that may grow or disappear. They are called **sunspots**. It is believed that these spots are areas where gases have broken through the surface. They are cooler than the surface of the Sun. On or between sunspots, there are powerful explosions lasting only a few moments. These explosions are known as **flares**. Huge sheets of glowing gases, called **prominences**, are seen leaping up from the Sun. They may reach 250,000 miles (400,000 km) into space.

Label these features on the photosphere, or surface, of the Sun on the drawing below.

prominences   sunspots   flares

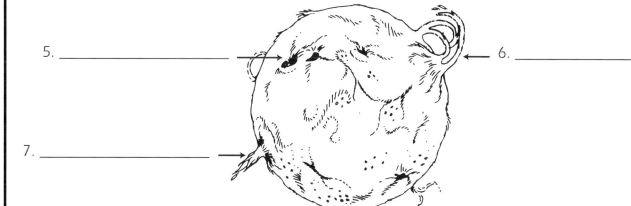

5. _____
6. _____
7. _____

Name _____   Date _____

SOLAR SYSTEM

REVIEW OF THE SUN

Write a short definition for each word. Then use the numbers to label the drawing below.

1. outer layer _____
2. middle layer _____
3. sunspot _____
4. flare _____
5. prominence _____
6. chromosphere _____
7. corona _____
8. core _____

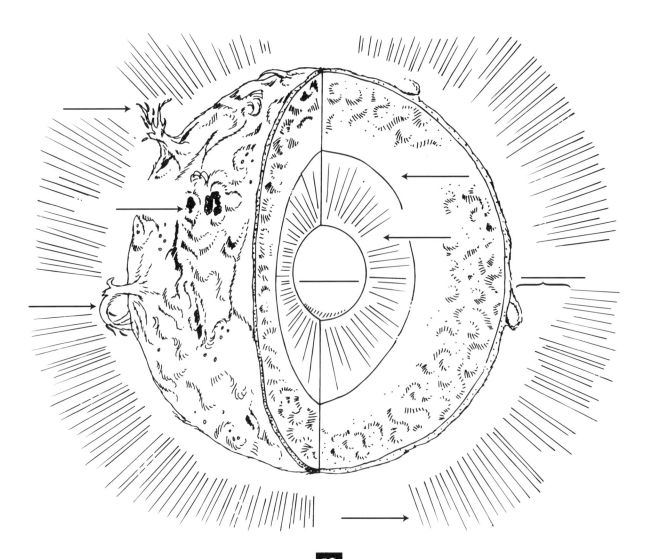

Name _____   Date _____

SOLAR SYSTEM

EARTH'S MOON

The moon is Earth's **satellite**. A satellite is an object that travels around a larger object. The moon travels around the Earth.

The Earth is about four times as large as the moon, our nearest neighbor in space. Because the moon is so close, it appears to be almost as big as the Sun. This is far from true.

The moon looks beautiful from Earth, but it is not a friendly place. Since the moon has such a weak gravitational pull, it has lost its atmosphere. Life, as we know it, cannot exist on the moon. The moon has no surface air or water. It is made up of hills, mountains, plains, and craters. There are no clouds, wind, or rain so the moon has no weather.

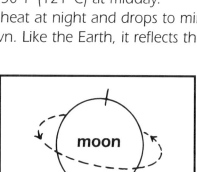

The moon does, however, have temperature. On the side of the moon facing the Sun, the temperature may rise to over 250°F (121°C) at midday. Then, because there is no air, the moon loses that heat at night and drops to minus 260°F (–162°C) or lower. The moon has no light of its own. Like the Earth, it reflects the light of the Sun.

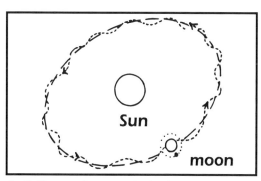

It revolves around the Earth.

It rotates on its axis

It follows the Earth in its movement around the Sun.

The moon always keeps the same side facing the Earth. Its period of rotation is the same as its period of revolution around the Earth—about 27 days.

SOLAR SYSTEM

PHASES OF THE MOON

Sun's Rays

New Moon

Waxing Crescent   Waning Crescent

Earth

First Quarter   Last Quarter

Gibbous Moon   Gibbous Moon

Full Moon

PHASES OF THE MOON

From our view on Earth, the moon seems to change shape from night to night. These changing shapes are called **phases**. The moon does not actually change. The shape we see depends on how much of the moon's lighted half we can see. Half of the moon is always lighted by the Sun, but we do not always see all of the bright side.

When the moon is between the Sun and Earth, its dark side is facing us. This is known as a new moon. It cannot be seen at all.

When the moon is on the far side of the Earth, and the Sun is on the opposite side of us, the moon is full. That is, the whole of the sunlit side is facing us and we see a full moon.

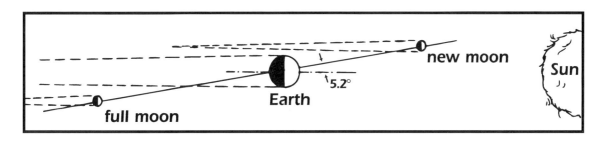

We have one full moon and one new moon each month. It takes about 27 days, 7 hours, and 43 minutes for the moon to complete one revolution of the Earth and to go through all phases. At different times the shape may be crescent, half, or gibbous (¾), depending on how much of the lighted half we can see at a particular time.

VIEWING PHASES OF THE MOON

MATERIALS
unshaded lamp   light colored ball

ACTIVITY

1. Darken the room and turn on the lamp. Hold the ball in front of you, in line with your eyes and the light bulb. The lamp is the Sun, the ball is the moon, and you are the Earth.

2. Begin moving the ball slightly to the left of the lamp. You will see a new moon. (The ball will appear completely dark.) Keep moving around the light with the ball in front of you. You will see each phase of the moon.

EXTRA CREDIT
Make a drawing showing the Sun, the Earth, and the moon in all its phases. Label each phase with its correct name.

Name _____ Date _____

SOLAR SYSTEM

REVIEW OF THE MOON

In each box, draw how the moon looks from Earth during its four main phases. The drawing shows how the moon would look from space.

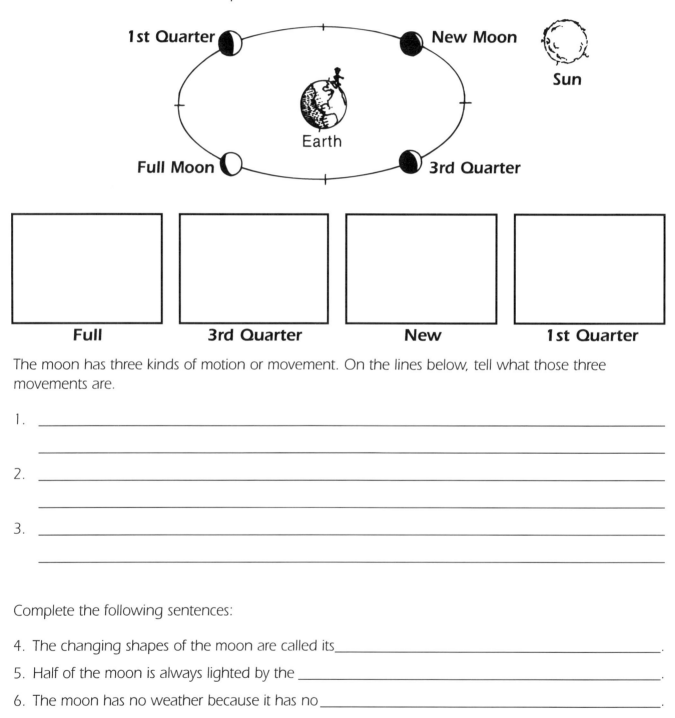

| Full | 3rd Quarter | New | 1st Quarter |

The moon has three kinds of motion or movement. On the lines below, tell what those three movements are.

1. _____
   _____

2. _____
   _____

3. _____
   _____

Complete the following sentences:

4. The changing shapes of the moon are called its _____.
5. Half of the moon is always lighted by the _____.
6. The moon has no weather because it has no _____.
7. The moon travels around the _____.
8. A _____ is an object that travels around another object.

SOLAR SYSTEM

INNER PLANETS

There are eight planets in our Solar System. They are dark spheres that reflect light from the Sun. They can be divided into two groups—inner planets and outer planets. The inner planets are Mercury, Venus, Earth, and Mars. The outer planets are Jupiter, Saturn, Uranus, and Neptune.

MERCURY

Mercury is the small planet closest to the Sun. Since it is between the Sun and Earth, it is often hidden in the Sun's glare. The Sun appears nine times larger on Mercury than on Earth. It bathes the planet in deadly radiation. Mercury is a ball of rock that has craters, hills, plains, and mountains. The days and nights on Mercury are long—the time between one sunrise and the next is 59 Earth days. Mercury is the speed demon of the Solar System, however, because it takes only 88 days to travel around the Sun.

How Old Would You Be on Mercury?

To keep track of your age on Mercury, you would simply have to remember that every 88 days you would be a year older—but a Mercurian year! How old would you be on Mercury? Figure out how many days old you are and divide that number by 88.

I am _____ days old on Earth and _____ years old on Mercury!

VENUS

Venus is second from the Sun and has an orbit twice as big as Mercury. Venus is sometimes called the morning or evening star because it appears shortly after sunset and before sunrise. With sunlight reflecting off its dense cloud cover, Venus is brighter than anything in the sky except for the Sun and moon. Because of its location between the Sun and Earth, Venus goes through phases as does our moon.

Venus is a hostile place. Its atmosphere is 98% carbon dioxide. The upper clouds are poisonous sulfuric acid. Its surface temperature is approximately 900°F (475°C). The atmosphere alone would crush you!

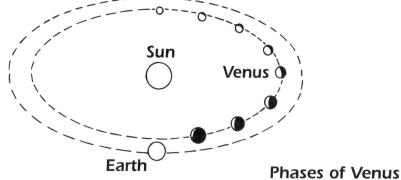

Phases of Venus

SOLAR SYSTEM

INNER PLANETS

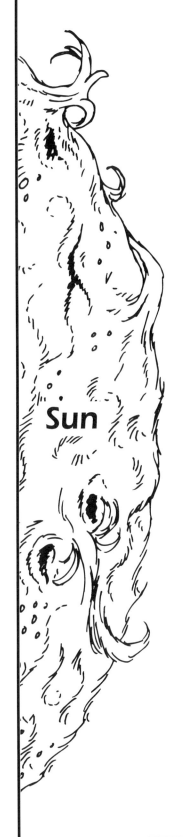
Sun

Mercury

Venus

Earth

Mars

SOLAR SYSTEM

INNER PLANETS

EARTH

Earth is the third planet from the Sun. It is a planet of intelligent life. Earth is a ball of rock and metal with a thin blanket of air. Much of the Earth's surface is covered with water. There are many reasons why this planet is perfect for life as we know it. The distance from the Sun is just right to receive plenty of heat and light, but not too much to bake in the Sun's radiation. Next, the orbit is nearly a circle which keeps the Earth always about the same distance from the Sun. This lets Earth get a constant steady flow of heat and light.

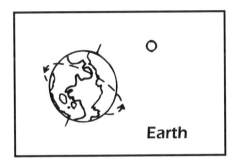
Earth

Because the Earth tilts and spins on its axis, it has gentle heating by day and cooling by night. Its atmosphere acts as a shield by day to filter out dangerous rays and as a blanket at night, to keep the heat from escaping into space. Earth's atmosphere is desirable for life.

Earth's atmosphere is made up of a combination of gases we can breathe—mainly nitrogen and oxygen. The size of our planet and the materials it is made of make our gravity just right to keep the atmosphere from escaping into space.

MARS

Mars is a small rocky planet with a thin atmosphere. It is fourth from the Sun and about half the size of Earth. A year on Mars is nearly twice as long as a year on Earth. It takes 687 Earth days for Mars to complete one orbit around the Sun. Mars has two moons or satellites, known as Phobos and Deimos.

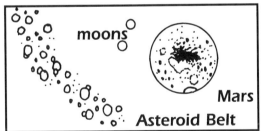
moons Mars Asteroid Belt

Mars is bone-dry on its surface and has giant volcanoes. Mars is tilted on its axis and has seasons, but they are twice as long as ours. Changes occur during the seasons. Martian ice caps grow in winter and shrink in summer. The average temperature by day is 86°F (30°C) and –103°F (–75°C) at night.

Just beyond Mars is a belt of tiny planets known as asteroids. They orbit the Sun as do the planets. Some asteroids are as big as mountains while others are quite small. Since they are like tiny planets, they are sometimes called planetoids.

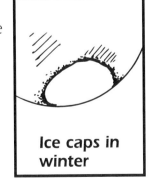
Ice caps in winter

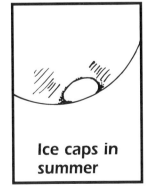
Ice caps in summer

SOLAR SYSTEM

REVIEW OF THE INNER PLANETS

Matching: Write the number of the statement that best fits each word.

A. _____ Mercury
B. _____ planets
C. _____ Earth
D. _____ atmosphere
E. _____ Venus
F. _____ asteroids
G. _____ Mars
H. _____ planetoids
I. _____ Jupiter
J. _____ phases

1. dark spheres that reflect light from the Sun
2. often referred to as the morning star
3. small planet containing giant volcanoes
4. planet closest to the Sun
5. contains much water and just the right amount of heat and light
6. the air surrounding a planet
7. tiny planets
8. another name for asteroids
9. an outer planet
10. the changes in shape that Venus appears to go through

Circle the words in the puzzle.

1. Find the names of the four inner planets.
2. Find these words:
   SUN
   ORBIT
   PHASE
   MOON
   STAR
   SOLAR
   ATMOSPHERE
   PLANET

```
B C S T A R F E
D M R U E Y A R
N O A A N T M E
P O M O O E G H
R N H Q R N V P
M E I C B A S S
X A U J I L T O
Z R K L T P V M
Y T S O L A R T
P H A S E U W A
```

Copyright © Milliken Publishing Co. All rights reserved.

MP3409 Solar System

Solar System

Outer Planets

Jupiter

Jupiter is the fifth planet from the Sun and the first of the four outer planets in our Solar System. It is the largest planet. You could fit eleven Earths along its diameter and more than a thousand Earths inside it. Jupiter is made up of twice as much material as all the other planets put together. Jupiter is so big and bright that you can see it from Earth without a telescope.

Jupiter takes almost 10 hours to spin around once on its axis and 12 Earth years to orbit the Sun once.

Jupiter is best known for its beautiful colors and its huge red spot. The colors make up the top of a clouded, churning atmosphere. The red spot is a giant, reddish, football-shaped mass believed to be a great storm system similar to a hurricane. The top of the atmosphere is a chilly −200°F (−128°C) or colder, while Jupiter itself is hot and believed to be made of gases and liquid metal.

Scientists now believe Jupiter has over 60 moons. The rings around Jupiter are composed of tiny particles.

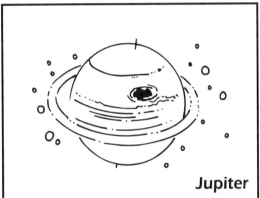
Jupiter

Extra Credit

Write a report about the satellites or moons of Jupiter. Include drawings if you wish.

Saturn

Saturn is a giant outer planet, sixth from the Sun. It is best known for the beautiful system of rings that circle the planet. The rings are made up of huge chunks of ice and tiny particles of dust and rock.

Saturn has a long orbit, taking 29½ Earth years to circle the Sun. Saturn is the second largest planet, but it is very light. So light, in fact, that it could float on water. Like Jupiter, Saturn is made mostly of gases. It also has over 60 known moons. Titan, the largest of Saturn's moons, is larger than Mercury.

Saturn

Extra Credit

Find out why Titan is a likely place to look for life.

Name _____   Date _____

SOLAR SYSTEM

OUTER PLANETS

Pluto*

Neptune

Uranus

Saturn

Jupiter

Mars

Mercury

Earth

Venus

Sun

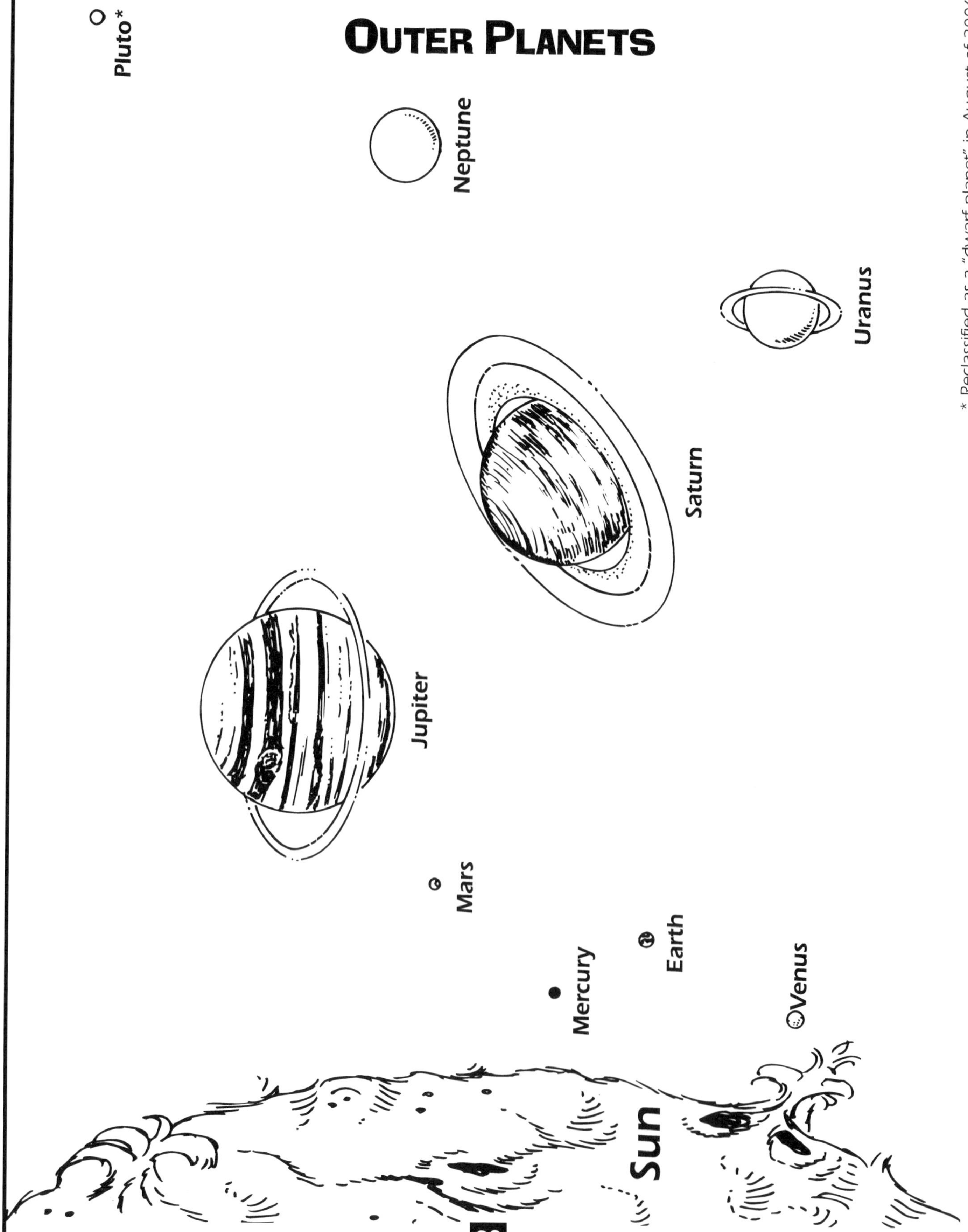

*Reclassified as a "dwarf planet" in August of 2006

20

Copyright © Milliken Publishing Co. All rights reserved.   MP3409 Solar System

SOLAR SYSTEM

OUTER PLANETS

URANUS

Uranus is the seventh planet from the Sun—a giant outer planet. Uranus is four times the size of Earth. Bluish-green in color, Uranus has 13 known rings with more than 25 moons orbiting the planet. Uranus lies on its side, unlike other planets.

A day on Uranus would last 23 hours and 15 minutes. That's how long it takes to spin around on its axis. However, it would take 84 years for Uranus to complete one orbit around the Sun. Deep within this planet lies an ocean which may be 6,000 miles deep.

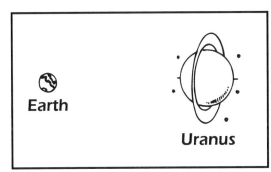

NEPTUNE

Neptune is eighth from the Sun and a large planet. It looks like Uranus because of its bluish color, but it is slightly smaller. Neptune can only be seen through a telescope. Even then, it is very difficult to study. It has at least 14 moons and several rings. Neptune takes 165 Earth years to orbit the Sun. Neptune contains a great dark spot as large as the planet Earth.

PLUTO

Until 2006, Pluto was considered the fifth outer planet. However, it was reclassified in a vote by the International Astronomical Union as a dwarf planet. Dwarf planets are defined as round objects that have not cleared the neighborhood around their orbits, and are not satellites (Pluto's orbit takes it through the Kuiper Belt, a large area of rocks at the edge of the solar system).

Pluto is so far out in space that it takes 248 Earth years to orbit the Sun once. Pluto has five known moons. Pluto's orbit is the shape of a long, flattened circle. Sometimes Pluto is closer to the Sun than Neptune. In fact, Pluto sometimes swings inside Neptune's orbit. It is very dark because it is so far out in space. Pluto was named after the god of the dead.

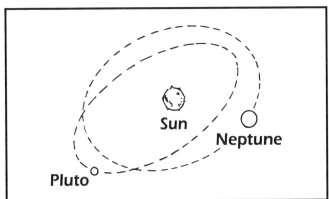

Astronomers believe that Pluto was once a moon of Neptune's. It was pulled out of orbit by something passing by. It is not at all like its giant neighbors with thick atmospheres. Pluto is more like the inner planets.

Name _____   Date _____

SOLAR SYSTEM

REVIEW OF OUTER PLANETS

Who am I? Use the clues to name the planet being described.

1. I am a bluish-green planet and have rings. Unlike other planets, I lie on my side. Who am I?   _____

2. I am the sixth planet from the Sun. I am known for my beautiful rings and my colors. I am made up mostly of gases. Who am I?   _____

3. I was reclassified as a "dwarf planet". Not much is known about me because I am so far away and so hard to study. I have five moons. Who am I?   _____

4. I am eighth from the Sun and have a greenish color. I can only be seen through a telescope. I have at least 14 moons. Who am I?   _____

5. I am the largest planet and well known for beautiful colors and my great red spot. I have rings and over 60 known moons. Who am I?   _____

SPACE MAIL

Pretend you are sending a letter to someone in outer space. It will be carried on board a spaceship. Fill in the information on the envelope below so the letter will reach your friend.

```
_____
_____
_____

                    Place
                    stamp
                    here

Name    _____
Street  _____
City    _____
State   _____
Country _____
Planet  _____
System  _____
Galaxy  _____
```

On a separate piece of paper, design a stamp to fit on the envelope above. Cut it out and paste it in the upper right corner.

SOLAR SYSTEM

Seasons

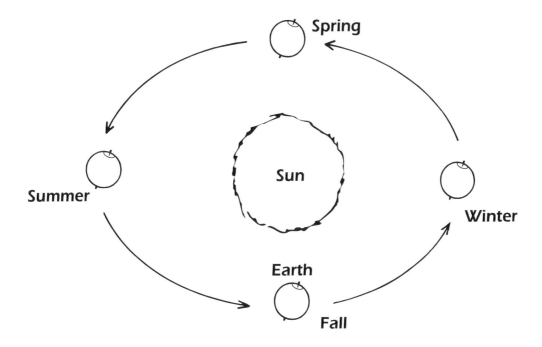

The Earth orbits the Sun. Earth is slightly tilted and spins on its axis. Because it is tilted and travels around the Sun, we have seasons. We go from summer, to fall, to winter, to spring, and then back to summer. These changes were most likely the first calendar people used. The time it took to complete these changes gave the length of time we call a year or 365¼ days. These changes were noted in two ways—change in temperature and change in the length of daylight.

We live in the Northern Hemisphere. It is summer when the North Pole is tilted toward the Sun. At this time, the Sun is high overhead and we receive strong, direct sun rays. The Sun shines for many hours each day. Its strong rays have a lot of time to heat the Earth. In the far north, it shines for 24 hours a day. This gradually changes—days get shorter and cooler and the Sun appears low in the sky at noon as the North Pole moves slowly away from the Sun. Summer turns to fall, and then to winter.

In winter, the North Pole is tilted away from the Sun. We do not receive the strong, direct rays and the Sun is low in the sky. The Sun shines for fewer hours each day. These weak rays do not have time to heat the Earth. This explains the colder winters even though the Sun is shining.

Winter turns to spring and then back to summer as the Earth completes one orbit around the Sun.

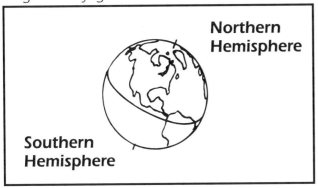

SOLAR SYSTEM

SEASONS IN THE NORTHERN HEMISPHERE

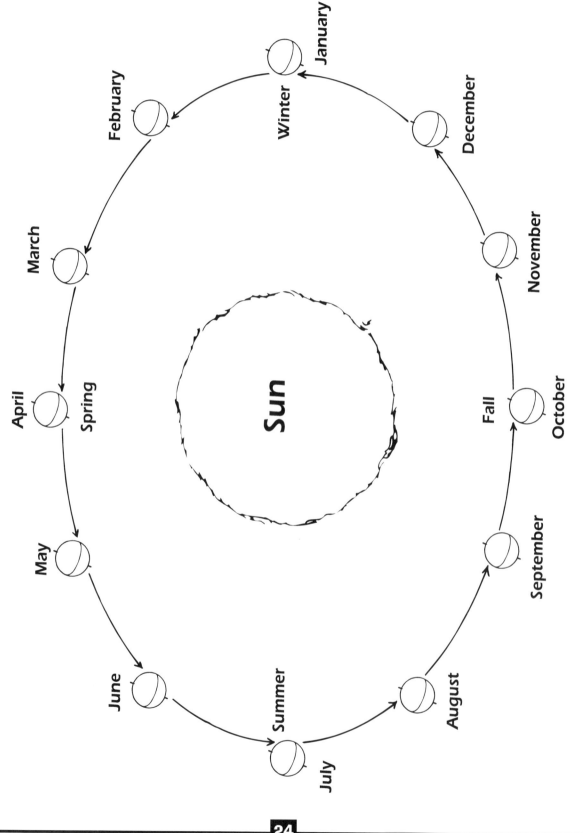

SOLAR SYSTEM

DAY AND NIGHT

Our days and nights are caused by the spinning of the Earth. Imagine the Earth is a ball of yarn with a knitting needle stuck through it. If you spin the needle, the ball will turn. We know there is nothing really poked through the center of the Earth, but the Earth still spins as if there were. This imaginary line through the center is called the axis. We say the Earth spins on its axis. It is at a slight tilt and spins from West to East. As the Earth rotates or spins on its axis, it is day when your part of the Earth is facing the Sun. It is night where you live when you are facing away from the Sun. The amount of time it takes for the Earth to spin around once on its axis is measured as one day—24 hours.

This is a view of the Earth from above the North Pole. Write the names of the times on the blanks.

Noon   Midnight
Sunset Sunrise

Complete the following:

1. The Earth spins on its _____.
2. The Earth revolves or travels around the _____.
3. It takes the Earth _____ hours to spin one complete time around.
4. When one side of the Earth is daytime, the opposite side is _____.
5. Explain in your own words why we have day and night. _____

EXTRA CREDIT

Tell what you believe would happen if the Earth suddenly stopped spinning and you were in the part of the world that had sunlight 24 hours a day. Include such things as how this would affect your sleep, work, school, and crime. You may include any other things of importance in your life and how they would be changed.

SOLAR SYSTEM

DAY AND NIGHT

EARTH TURNING ON ITS AXIS

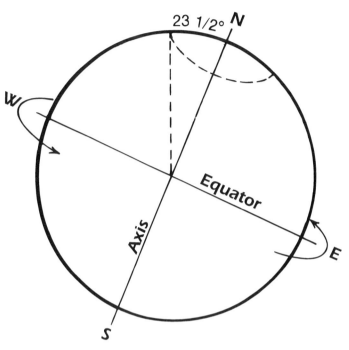

DAY AND NIGHT

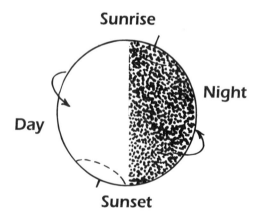

It takes 24 hours for the Earth to complete one rotation on its axis.

SOLAR SYSTEM

REVIEW OF THE SEASONS

Fill in the blanks below with the correct vowels to make the sentences true. Remember, the vowels are a, e, i, o, u, and y.

1. W __ h __ v __ s __ __ s __ ns b __ c __ __ s __ th __ __ __ rth __ s t __ lt __ d __ nd tr __ v __ ls __ r __ __ nd th __ S __ n.

2. Th __ f __ __ r s __ __ s __ ns __ r __: s __ mm __ r, w __ nt __ r, f __ ll, __ nd spr __ ng.

3. W __ l __ v __ __ n th __ N __ rth __ rn H __ m __ sph __ r __.

4. __ t __ s s __ mm __ r wh __ n th __ N __ rth P __ l __ __ s t __ lt __ d t __ w __ rd th __ S __ n.

5. Th __ __ __ rth __ rb __ ts th __ S __ n __ n 365¼ d __ ys.

6. __ n w __ nt __ r, th __ N __ rth P __ l __ __ s t __ lt __ d __ w __ y fr __ m th __ S __ n.

7. __ n th __ s __ mm __ r, th __ S __ n __ s h __ gh __ v __ rh __ __ d __ nd w __ r __ c __ __ v __ str __ ng, d __ r __ ct r __ ys.

8. Th __ S __ n __ s l __ w __ n th __ sk __ __ t n __ __ n d __ r __ ng th __ w __ nt __ r.

9. Th __ __ __ rth sp __ ns __ n __ ts __ x __ s.

10. __ n w __ nt __ r, w __ r __ c __ __ v __ f __ w __ r h __ __ rs __ f s __ nl __ ght th __ n w __ d __ __ n th __ s __ mm __ r.

Solar System Crossword Puzzle Review

Down

1. Planet of giant volcanoes
2. Center of our Solar System
4. We hear with this part of the body.
5. One orbit around the Sun is called a _____.
6. Everything in space is a part of the _____.
8. A tiny particle
9. The Earth spins on its _____.
10. Center of the Sun
12. Reclassified as a dwarf planet
13. It gives off its own heat and light.
14. Planet with a beautiful ring system
15. Another name for automobile
18. We live in the Milky Way _____.
19. The largest planet
21. Changes in temperature and length of days
23. Changes in shape the moon seems to go through
24. Eighth planet from the Sun
27. Uranus is one of the four _____ Planets.

Across

3. Planet closest to the Sun
6. Planet that lies on its side
7. Planet we live on
9. "Tiny planets" that orbit between Mars and Jupiter
11. Outer part of the Sun's atmosphere
13. The Sun is the center of the _____.
15. Loose collection of rock and frozen gas that has a tail
16. 2,000 pounds = 1_____
17. The Earth's satellite
20. The Earth _____ on its axis.
22. To rotate means to _____.
23. Asteroids are often called _____.
25. Evening or morning "star"
26. Dark, cool spot on the Sun
28. The Sun gives off heat and light _____.
29. Round sphere that does not give off light of its own

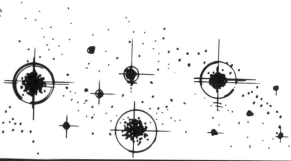

SOLAR SYSTEM BACKGROUND MATERIAL

The study of our Solar System is fascinating, but ever-changing. As man probes further into space, new discoveries are made which replace old theories. The information in this book should be used as a springboard for study and observation to enhance new discoveries as they are made.

Pages 1–3: The members of the Solar System are held in orbit by the gravitational force of the Sun. These orbits are very stable and have existed for billions of years.

There is a belt of asteroids between Mars and Jupiter. Thousands of large rocks and small pebble-like particles make up this belt and orbit the Sun. Scientists theorize that asteroids may be the remains of a planet that burst apart; pieces of material that did not come together to form a planet; or "dead" comets.

The smallest objects are meteors, which occasionally stray into the Earth's atmosphere and glow as a falling star.

Comets have a bright head and a long, shining tail. They shine partly because of light absorbed and reradiated from the Sun and partly because of reflected sunlight. Their orbits are elliptical and may be on different planes than that of the planets. When approaching the Sun, they develop long tails that always point away from the Sun.

The stars we see at night are not part of our Solar System. They are millions of miles away. The star nearest to our Sun is 25 million miles away. There are many other stars in the universe that may have systems similar to the one in which we live.

Students may better understand the size of planets and their relative distance from the Sun by constructing a model of the Solar System. Students will need to research the distance each planet is from the Sun and establish a scale to determine the size they will make their planets.

Planets may be made from papier mâché. Larger planets may be molded over blown-up balloons. Models could be suspended from the ceiling of the classroom and labeled. Moons and other bodies may be added.

Pages 4–5: Two orbiting bodies have a common center of gravity, always located between them, but nearer to the center of the system. In the Earth-moon system, the center of gravity lies under the crust of the Earth.

Bodies that orbit are called satellites. There are two kinds—natural and artificial. The moon and the planets are natural satellites. The space age has brought about the launching of hundreds of artificial satellites.

For extra credit: In the late 1500s, a famous astronomer, Johannes Kepler, determined several important laws of motion to explain and predict the motions of planets. Before this time, it was believed that planets orbited the Sun in perfect circles. Kepler mathematically showed their orbits to be ellipses.

Pages 7–9: The Sun is made up mostly of hydrogen that produces helium under pressure and causes energy to be released. The Sun is its own fuel. It uses up 600 million tons of hydrogen each second. It is estimated that, at that rate, it can keep shining for at least 6 billion more years. The Earth receives a very small portion of the Sun's energy.

Pages 11–13: Many questions about the moon were answered when Neil Armstrong first stepped out of Apollo 11 and onto the moon on July 20, 1969.

The moon goes through phases as it orbits the Earth. One half of the moon is always lighted by the Sun, but the phases are the shapes of the lighted moon that we see from Earth. Moon's phases go from full moon to gibbous to last quarter. From last quarter, the phases go to crescent and back to new moon. The moon is said to be waxing when it is showing more and more of its face. It is waning when it is showing less and less of its face.

Our nearest neighbor in space, the moon, is the Earth's only natural satellite. Because the moon's gravity is so weak, there is no atmosphere there and the moon is completely silent.

Demonstrate rotation and revolution by having two students stand facing each other. Student A (Earth) stands in the same spot while student B (moon) walks around him. They should always stay face to face. A will turn in place while B walks around him.

Pages 15–21: The planets are small, dark spheres. They have no light of their own, but appear to glow at times because they reflect light from the Sun. We know of eight major planets (Pluto, formerly the ninth planet, was reclassified as a "dwarf planet" in August of 2006). The planets are divided into two categories: inner planets and outer planets.

The inner planets are small and composed of a rocky material. They are wrapped in an atmosphere of gas (hydrogen and helium).

The planets are quite different from each other in many ways. Students should find additional research on the individual planets very interesting, particularly in light of recent discoveries.

For extra credit: Titan is the only known moon to have a thick atmosphere.

Pages 23–27: As different amounts of heat from the Sun are received, seasonal changes occur. Because of the tilt of the Earth on its axis, the North Pole tilts more toward the Sun in summer and slightly away from the Sun in winter. As the Earth orbits the Sun, the rays of the Sun strike the atmosphere continually changing angles. The more vertical the rays are, the more heat an area receives. Slanting rays are weaker because they spread out the same solar energy over a larger area. The rays also go through more atmosphere which reflects some heat to outer space before the rays reach Earth's surface. Another reason for seasonal changes is that the days are longer in summer. An area receives direct rays for more hours a day.

The poles have extremes of the seasons. The rays of the Sun are extremely slanted, and there are periods of continuous darkness in winter. Places along the equator do not have seasonal changes because the Sun's rays are nearly vertical all year long.

Seasons in the Southern Hemisphere occur in reverse order from those in the Northern Hemisphere.

Demonstrate direct and vertical rays with the students by holding one flashlight pointing straight down toward the floor and a second flashlight pointed out at the floor in front of you. Slanting rays spread out over a large area, while vertical rays are direct, strong rays.

As the Earth spins on its axis, it is day when the part of the Earth where you live is facing the Sun. It is night when the area where you live is away from the Sun.

Some people use a 24-hour clock. The military uses this system of telling time. The hours are numbered from 0:00 to 23:59. Midnight is zero hour. After 12 noon, which is 1200, comes 1300, 1400, etc. By this system, students may get out of school around fifteen hundred and dinner might be around eighteen hundred. Ask students questions requiring answers in 24-hour time.

Additional Activities for a Solar System unit: Have students research and report on any of the following people: Percival Lowell, Galileo, Robert H. Goddard, Johann Bode, Wm. Herschel, Isaac Newton, Nicolaus Copernicus, Christiaan Huygens, Edwin Hubble, Hans Lippershey, Edmund Halley, or Clyde W. Tombaugh.

Make a planet wall poster using a piece of string about 30 inches long. Pin the string's end to the middle of a strip of paper about 30 inches long. Mark the string at intervals to show each planet in its correct position from the Sun.

Observing the moon phases is a month's project. Have students start when there is a new moon and observe the moon every other night. Have them begin by putting each date the moon is to be observed on the left side of a 3" x 5" index card. On the right side of the card, have them lightly draw a circle two inches in diameter. Every night when the moon is observed, they can draw the shape of the moon in the circle on the card. If it is cloudy one night, observe the next night. When finished, have them put all the cards in order, hold on the left side and flip the cards to see the phases of the moon.

SOLAR SYSTEM

ANSWERS

Page 3
1. solar
2. loose rock and frozen gas
3. A planet does not give off light, it reflects the light of the Sun. Stars produce and give off their own heat and light.
4. Sun
5. The planets revolve around the Sun.
6. Milky Way
7. Our Solar System is a tiny speck.

Page 6
1. F A planet reflects the light of the Sun.
2. T
3. F The Universe is made up of everything in space.
4. F The moon orbits the Earth.
5. F Neptune has the slowest orbit out of the eight planets. Mercury the fastest.
6. T
7. F The Earth takes one year to orbit the Sun.
8. T
9. T Pluto, formerly the ninth planet, was reclassified as a "dwarf planet" in August, 2006.
10. F Mercury, Venus, Earth, Mars, Jupiter, Saturn, Uranus, Neptune
11. T

Page 7
A. corona
B. chromosphere
C. outer layer
D. middle layer
E. core

Page 9
1. Energy is a force.
2. An atom is a tiny particle.
3. Hydrogen is a gas.
4. Helium is a gas.
5. sunspot
6. prominence
7. flare

Page 10
1. layer closest to the surface of the Sun
2. layer next to the core of the Sun
3. dark spots on the surface of the Sun, caused by cool gases
4. powerful explosions on the Sun
5. huge sheets of glowing gases that leap from the Sun
6. inner part of the Sun's atmosphere
7. outer part of the Sun's atmosphere
8. center of the Sun

Page 14
1. The moon revolves around the Earth.
2. The moon rotates on its axis.
3. The moon follows the Earth in its movement around the Sun.
4. phases
5. Sun
6. atmosphere
7. Earth
8. satellite

Page 18
A. 4
B. 1
C. 5
D. 6
E. 2
F. 7
G. 3
H. 8
I. 9
J. 10

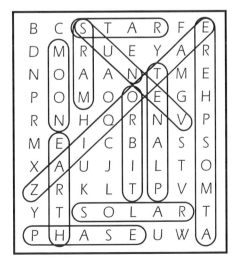

SOLAR SYSTEM

Page 22
1. Uranus
2. Saturn
3. Pluto
4. Neptune
5. Jupiter

Page 25
1. axis
2. Sun
3. 24
4. night
5. The Earth spins on its axis. Spinning causes half of the Earth to be in sunlight and the opposite side to be in darkness. It takes 24 hours for this to happen.

Page 27
1. We have seasons because the Earth is tilted and travels around the Sun.
2. The four seasons are: summer, winter, fall, and spring.
3. We live in the Northern Hemisphere.
4. It is summer when the North Pole is tilted toward the Sun.
5. The Earth orbits the Sun in 365 1/4 days.
6. In winter, the North Pole is tilted away from the Sun.
7. In the summer, the Sun is high overhead and we receive strong, direct rays.
8. The Sun is low in the sky at noon during winter.
9. The Earth spins on its axis.
10. In winter, we receive fewer hours of sunlight than we do in the summer.

Page 28

Down
1. Mars
2. Sun
4. ear
5. revolution
6. universe
8. atom
9. axis
10. core
12. Pluto
13. star
14. Saturn
15. car
18. galaxy
19. Jupiter
21. seasons
23. phases
24. Neptune
27. outer

Across
3. Mercury
6. Uranus
7. Earth
9. asteroids
11. corona
13. Solar System
15. comet
16. ton
17. moon
20. rotates
22. turn
23. planetoids
25. Venus
26. sunspot
28. energy
29. planet